3 Minnesota MCA Grade 4 Math Practice Tests

Full-Length Test Prep with Detailed Answer Explanations

Dr. A. Nazari

Grade 4 Math: 3 Practice Tests

Hey there, quick checker! 

Three tests. That's all it takes to **see where you stand** and **build your confidence** fast.

✅ Short and focused — perfect for a quick review.

✅ Find your strengths in just three rounds.

✅ Fast results, real progress!

Ready for a quick check? Let's go! 👍

> 66 *Three quick tests and you'll know exactly what to study next!* 99

📋 How to Use This Book 📋

📖 What's Inside

- **3 Practice Tests** — quick, focused tests covering all Grade 4 topics
- **Answer Key** — detailed answers and explanations at the back
- **Reference Pages** — math symbols and multiplication table you can use during tests
- **My Test Tracker** — record your scores and see your growth at a glance

📋 How to Take a Practice Test

1. **Set up:** Find a quiet spot. Gather pencils, eraser, and scratch paper.
2. **Time it:** Set a timer (or ask a grown-up to). Try to finish without rushing.
3. **Work through it:** Answer every question. Skip hard ones and come back later.
4. **Check your work:** Use any extra time to double-check your answers.
5. **Score it:** Use the Answer Key. Write your score in the Test Tracker.

☑ Multiple Choice

Pick the best answer from the choices given. Read ALL options before choosing — the first one that looks right isn't always the best!

✏ Short Answer

Solve the problem and write your answer. Show your work — even if you get the answer wrong, partial credit counts in many real tests!

ⓘ Quick Tip: With only 3 tests, make each one count. Review every missed question before moving to the next test!

Test-Taking Tips

Quick strategies to help you ace all 3 tests!

Before You Start

- Get a good night's sleep and eat a healthy snack.
- Have your supplies ready: pencils, eraser, scratch paper.
- Take three slow, deep breaths — you've got this!
- Remember: just 3 tests, so give each one your full focus.

During the Test

1. **Read each question twice.** Underline key words like "how many more," "product," or "estimate."
2. **Show your work.** Write out each step — it helps you catch mistakes.
3. **Use scratch paper.** Line up digits carefully for multi-digit problems.
4. **Skip and return.** Stuck on a question? Star it and move on. Come back with fresh eyes.
5. **Check with estimation.** Does your answer make sense? A quick estimate can catch big errors.
6. **Use your time wisely.** Don't spend too long on one problem.

Multiple Choice Tricks

- Read **all** choices first.
- Cross out answers you know are wrong.
- Plug your answer back into the problem.
- When in doubt, eliminate and guess — never leave it blank!

Watch Out For...

- Rushing through without reading carefully.
- Mixing up $\times$ and $+$ in word problems.
- Forgetting to regroup when adding or subtracting.
- Not simplifying fractions when asked.
- Skipping the "check your work" step.

"Three tests is all you need for a quick check-up! Focus on each one, learn from your mistakes, and you'll be amazed how much you grow."

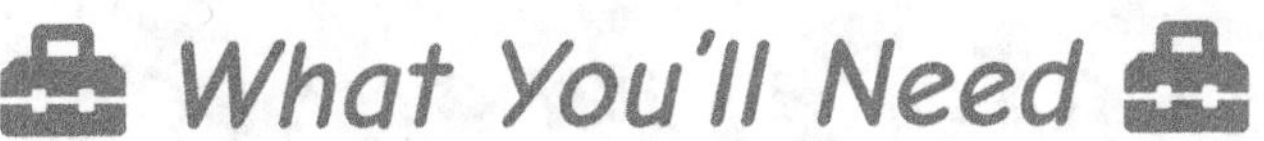# What You'll Need

Grab these supplies before you start each test.

Pencils

#2 pencils are perfect!

Eraser

Clean erasing = cleaner work

Scratch Paper

For all your work!

Quiet Space

Like a real test room

Timer (Optional)

Practice pacing yourself

Confidence!

Three tests, three wins!

✓ You CAN Use

- Pencils and eraser
- Scratch paper
- The reference pages in this book
- A ruler (for measurement questions)

✗ You CANNOT Use

- Calculator
- Phone or tablet
- Other books or notes
- Help from anyone else

👥 For Parents & Teachers

*This compact 3-test set is ideal for a quick skills check. Simulate real test conditions when possible. After each test, review missed questions together and focus on **growth** ("You improved on fractions this time!") rather than the raw score.*

X^1 Math Reference Sheet X^1

You may use this page during your practice tests!

Symbol	Name	Meaning
$+$	Plus	Combine amounts
$-$	Minus	Find the difference
$\times$	Times	Multiply (repeated groups)
$\div$	Divide	Split into equal parts
$=$	Equals	Same value on both sides
$>$ $<$	Greater / Less Than	Compares two values
$\frac{a}{b}$	Fraction	a parts out of b equal parts
$.$	Decimal Point	Separates wholes from parts
$\angle$	Angle	Measured in degrees

- **Factor** — a number you multiply
- **Product** — answer from multiplying
- **Quotient** — answer from dividing
- **Remainder** — left over after dividing
- **Numerator** — top of a fraction
- **Denominator** — bottom of a fraction
- **Equivalent** — equal in value
- **Perimeter** — distance around
- **Area** — space inside
- **Estimate** — a close, rounded guess

🔍 Word Problem Clue Words

Add ($+$)   in all, altogether, total, combined, sum, increase

Subtract ($-$)   how many more, how many fewer, difference, left, remain

Multiply ($\times$)   each, every, per, times as many, groups of, product

Divide ($\div$)   split equally, shared among, divided into, per group

◼ Multiplication Table ◼

You may use this table during your practice tests!

×	1	2	3	4	5	6	7	8	9	10	11	12
1	1	2	3	4	5	6	7	8	9	10	11	12
2	2	4	6	8	10	12	14	16	18	20	22	24
3	3	6	9	12	15	18	21	24	27	30	33	36
4	4	8	12	16	20	24	28	32	36	40	44	48
5	5	10	15	20	25	30	35	40	45	50	55	60
6	6	12	18	24	30	36	42	48	54	60	66	72
7	7	14	21	28	35	42	49	56	63	70	77	84
8	8	16	24	32	40	48	56	64	72	80	88	96
9	9	18	27	36	45	54	63	72	81	90	99	108
10	10	20	30	40	50	60	70	80	90	100	110	120
11	11	22	33	44	55	66	77	88	99	110	121	132
12	12	24	36	48	60	72	84	96	108	120	132	144

💡 Quick Reminders

Multiply: Find one number on the left, the other on top. Where they meet = your answer.

Divide: For 96 ÷ 8, find 96 in the 8-row. The column header tells you the answer: 12!

Factor pairs: Any number in the table can be written as row header × column header.

Name: _________________________________

Test #	Date	Score	How I Feel
1			
2			
3			

Three tests, three chances to shine! Compare each score to your last one — even one extra point means you learned something new!

Quick progress is still progress! Three tests gave you a clear picture. Now you know exactly where to focus next. Be proud of every step forward!

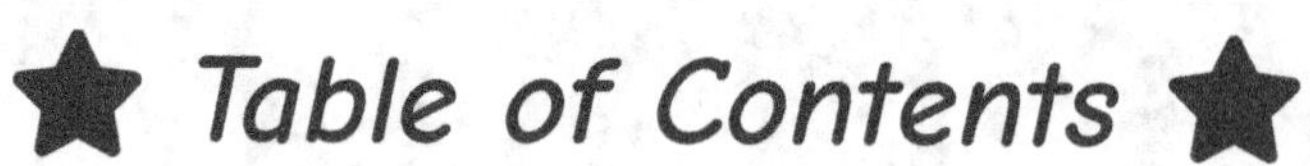

★ Table of Contents ★

Here's what we'll explore together!

 Let's learn and have fun!

1

Practice Test 1

 30 Questions

✏ Before You Start ✏

- ✔ **Read each question carefully** before choosing your answer.
- ✔ **Show your work** on scratch paper when you need to.
- ✔ **Skip hard questions** and come back to them later.
- ✔ **Check your answers** when you're done.
- ✔ **Take your time** — there's no rush!

⭐ You've Got This! ⭐

Do your best and show what you know!

1. A box has 5 pencils. Another box has 7 times as many pencils. How many pencils are in the bigger box?

Your Answer:

2. There are 45 students split equally into 5 teams. Each team needs 3 jump ropes. How many jump ropes are needed in total?

Your Answer:

3. Is 100 a multiple of 4? Is 100 a multiple of 6? Answer both questions.

Your Answer:

4. Find the rule and fill in the missing value in the table:

In	Out
3	15
4	20
6	30
9	?

Your Answer:

5. In 307,416, which digit is in the hundred-thousands place?

(A) 7

(B) 4

(C) 3

(D) 0

6. Which comparison is **INCORRECT**?

   (A) $745,000 > 74,500$

   (B) $600,001 > 599,999$

   (C) $410,300 < 410,030$

   (D) $325,400 = 325,400$

7. If $3 \times 10 = 30$, what is 3×100?

   (A) 30

   (B) 300

   (C) 3,000

   (D) 3,030

8. Which is the best estimate for $387 + 425$ using rounding to the nearest hundred?

   (A) 700

   (B) 800

   (C) 900

   (D) 750

9. List the partial products for 7×428 and find the total.

   Your Answer:

10. You have $47 to spend on notebooks that cost $5 each. How many notebooks can you buy?

    (A) 8 notebooks

    (B) 9 notebooks

    (C) 10 notebooks

    (D) 11 notebooks

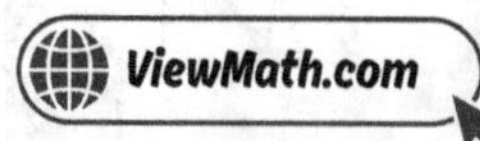

11. Ben ate 3 slices of a pizza cut into 6 equal slices. Each slice is $\frac{1}{6}$ of the pizza. How much pizza did Ben eat?

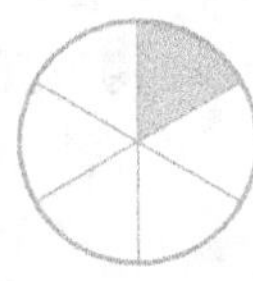

(A) $\frac{1}{6}$

(B) $\frac{2}{6}$

(C) $\frac{3}{6}$

(D) $\frac{6}{3}$

12. Maya had $\frac{9}{10}$ of a tank of gas. She drove to school (using $\frac{2}{10}$) and to soccer practice (using $\frac{3}{10}$). How much gas is left?

(A) $\frac{6}{10}$

(B) $\frac{5}{10}$

(C) $\frac{4}{10}$

(D) $\frac{7}{10}$

13. What is $3\frac{4}{5} + 2\frac{3}{5}$? (Remember to regroup if the fraction part is improper!)

(A) $5\frac{7}{5}$

(B) $5\frac{2}{5}$

(C) $6\frac{2}{5}$

(D) $6\frac{7}{5}$

14. What is $5 \times \frac{2}{8}$, written as a mixed number?

(A) $1\frac{1}{8}$

(B) $1\frac{3}{8}$

(C) $1\frac{2}{8}$

(D) $\frac{10}{8}$

15. A student says $\frac{1}{10} + \frac{3}{100} = \frac{4}{110}$. What is the student's mistake?

  (A) The student added the denominators instead of finding a common denominator.

  (B) The student forgot to add the numerators.

  (C) The student should have multiplied the denominators.

  (D) There is no mistake.

16. 0.6 can also be written as a fraction with a denominator of 100. Which fraction is it?

  (A) $\frac{6}{100}$

  (B) $\frac{60}{100}$

  (C) $\frac{16}{100}$

  (D) $\frac{600}{100}$

17. Compare: 0.34 ___ 0.43

  (A) $>$

  (B) $<$

  (C) $=$

  (D) Cannot be determined.

18. Which conversion is correct?

  (A) $1\ cm = 10\ mm$

  (B) $1\ m = 10\ cm$

  (C) $1\ km = 100\ m$

  (D) $1\ kg = 100\ g$

19. How many inches are in 1 foot?

  (A) 10

  (B) 12

  (C) 16

  (D) 3

20. Look at the diagram below. A track is 400 m around. The runner has completed 3 laps. How many kilometers has the runner covered?

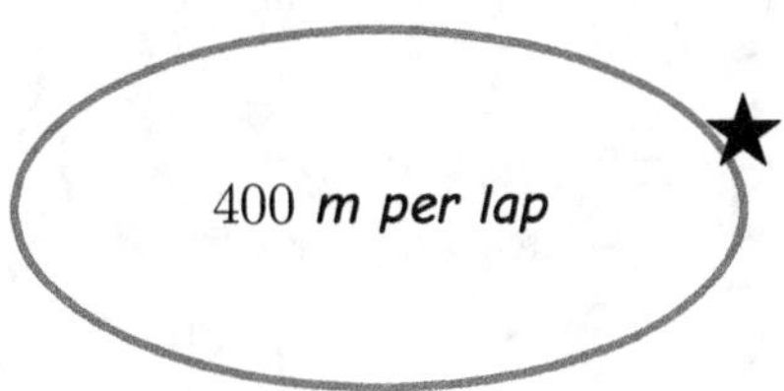

Laps completed: 3

Your Answer:

21. A composite shape is a large rectangle 8×6 m with a 2×2 m square cut out of one corner. What is the remaining area?

(A) 40 sq m

(B) 44 sq m

(C) 46 sq m

(D) 48 sq m

22. A rectangle has perimeter 36 cm and length 10 cm. What is its width?

(A) 6 cm

(B) 8 cm

(C) 10 cm

(D) 16 cm

23. A number line for a line plot measures in $\frac{1}{2}$-cup intervals. Which values should appear on the number line between 0 and 1?

(A) $\frac{1}{4}$

(B) $\frac{3}{4}$

(C) $\frac{1}{2}$

(D) $\frac{2}{3}$

24. Which angle measure is obtuse?

   (A) 45°

   (B) 89°

   (C) 110°

   (D) 180°

25. Angle A measures 45° and Angle B measures 45°. Together, what type of angle do they form?

   (A) Acute

   (B) Right

   (C) Obtuse

   (D) Straight

26. Points C, D, and E are all on the same line. How is segment $\overline{CD}$ related to segment $\overline{DE}$?

   (A) They are the same segment

   (B) They are parts of the same line

   (C) They are parallel

   (D) They are perpendicular

27. Which angle type has the LARGEST measure of the four types?

   (A) Acute

   (B) Right

   (C) Obtuse

   (D) Straight

28. True or false: Parallel lines can intersect if they are made long enough.

   Your Answer:

29. Which statement is ALWAYS true about a rectangle?

   (A) All four sides are equal

   (B) It has exactly two right angles

   (C) It has four right angles

   (D) It is not a parallelogram

30. *How many lines of symmetry does an equilateral triangle (all sides equal) have?*

(A) 1

(B) 2

(C) 3

(D) 0

Find more at
ViewMath.com/MN-Grade4

 End of Practice Test 1 

Great job finishing the test!

📋 My Score

I got __________ out of 30 questions right.

*Check your answers in the **Answer Key** at the back of the book.*

💡 *Review any questions you missed. That's how we learn!*

📊 Check Your Score Online!

Visit **ViewMath Academy** to enter your answers and see which topics you need to review. You can also explore lessons, take quizzes, track your scores, and save your progress!

viewmath.com/score/4.1.MN.01

Or go to viewmath.com/score and enter code: 4.1.MN.01

2

Practice Test 2

 30 Questions

✏ Before You Start ✏

- ✔ **Read each question carefully** before choosing your answer.
- ✔ **Show your work** on scratch paper when you need to.
- ✔ **Skip hard questions** and come back to them later.
- ✔ **Check your answers** when you're done.
- ✔ **Take your time** — there's no rush!

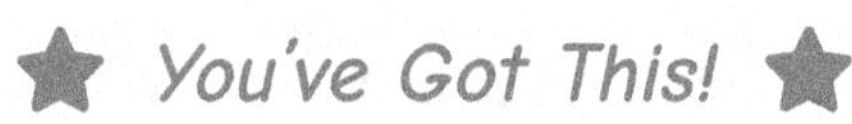 ⭐ You've Got This! ⭐

Do your best and show what you know!

1. How many total dots are in the array below? Write a multiplication equation for the array.

9 columns

3 rows

Your Answer:

2. A school ordered 7 boxes of crayons. Each box has 24 crayons. The teacher gave away 15 crayons. How many crayons does the school have now?

(A) 168

(B) 153

(C) 163

(D) 152

3. I am a prime number between 45 and 55. What am I? (There may be more than one answer.)

Your Answer:

4. A pattern starts at 3 and the rule is "multiply by 3." What is the fourth term?

(A) 12

(B) 27

(C) 81

(D) 9

5. What is the value of the digit 5 in 850,293?

(A) 5

(B) 500

(C) 5,000

(D) 50,000

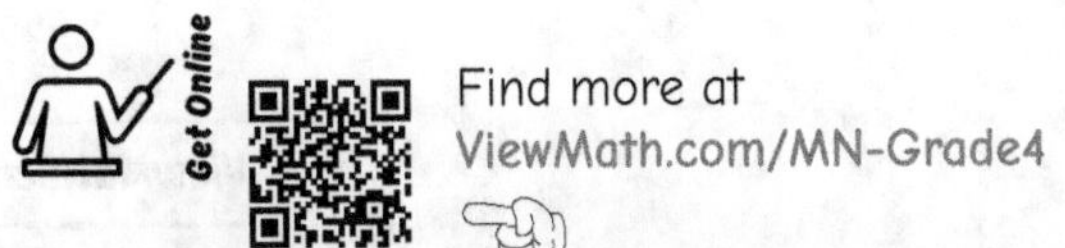

6. The population of Town A is 183,400 and Town B is 183,040. Which town has more people?

(A) Town B

(B) They are equal

(C) Town A

(D) Cannot tell

7. Look at the diagram below. The same digit 9 appears in two positions. How many times bigger is the value at Position A than at Position B?

TTh	Th	H	T	O
9	2	9	4	1

↓   ↓
A   B

Your Answer:

8. Round 347 to the nearest ten.

(A) 300

(B) 340

(C) 350

(D) 400

9. Each page of a photo album holds 7 photos. The album has 215 pages. How many photos can the album hold?

Your Answer:

10. What is $504 \div 7$?

(A) 70

(B) 71

(C) 72

(D) 73

Find more at
ViewMath.com/MN-Grade4

11. *Find the missing numerator.* $\frac{6}{8} = \frac{?}{8} + \frac{2}{8} + \frac{1}{8}$

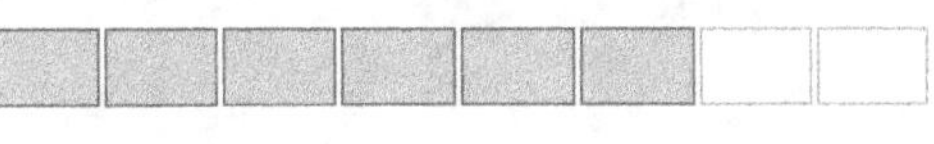

12. *What is* $\frac{6}{8} - \frac{2}{8}$?

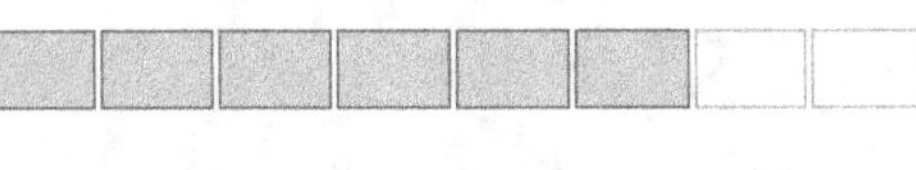

(A) $\frac{1}{2}$

(B) $\frac{3}{8}$

(C) $\frac{4}{8}$

(D) $\frac{5}{8}$

13. *Convert* $3\frac{2}{5}$ *to an improper fraction.*

(A) $\frac{11}{5}$

(B) $\frac{15}{5}$

(C) $\frac{17}{5}$

(D) $\frac{16}{5}$

14. *A student says* $3 \times \frac{2}{5} = \frac{6}{15}$. *What mistake did the student make?*

(A) *The student forgot to add rather than multiply.*

(B) *The student multiplied the denominator instead of keeping it the same.*

(C) *The student multiplied the numerator correctly.*

(D) *There is no mistake;* $\frac{6}{15}$ *is correct.*

15. What is $\frac{3}{10} + \frac{5}{100}$?

(A) $\frac{8}{100}$

(B) $\frac{8}{110}$

(C) $\frac{35}{100}$

(D) $\frac{35}{110}$

16. Are 0.30 and 0.3 equal?

(A) No, because 0.30 has two decimal places.

(B) No, because 0.30 is bigger than 0.3.

(C) Yes, adding a zero at the end of a decimal does not change its value.

(D) Yes, but only when the denominator is 100.

17. Look at the two 10×10 grids below. Which decimal is greater?

Grid A: 0.42

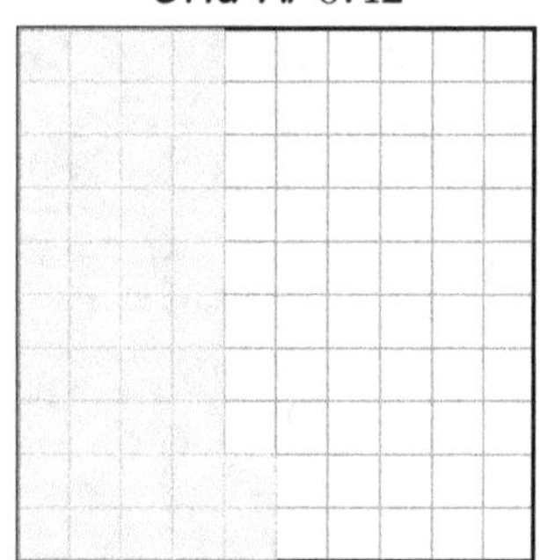

Grid B: 0.38

(A) 0.42 is greater

(B) 0.38 is greater

(C) They are equal

(D) Cannot tell from the grids

18. How many centimeters are in 2 kilometers? (two-step: km → m → cm)

(A) 2,000 cm

(B) 20,000 cm

(C) 200,000 cm

(D) 2,000,000 cm

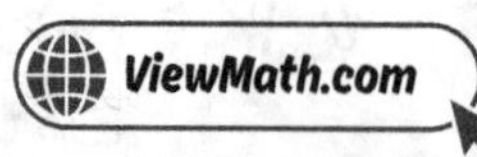

19. Look at the balance scale. The left side has 5 pounds. How many ounces are needed on the right to balance it?

(A) 60 oz

(B) 80 oz

(C) 50 oz

(D) 90 oz

20. Look at the bar model below. Find the total weight of both packages in grams.

Total weight = ?	
Package A: 4 kg	Package B: 1,500 g

Your Answer

21. A room is 9 ft wide. Its area is 108 sq ft. How long is the room?

(A) 10 ft

(B) 11 ft

(C) 12 ft

(D) 13 ft

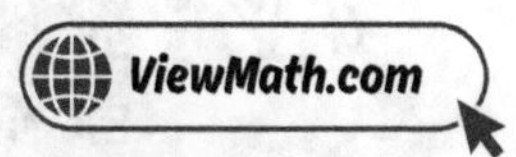

22. Look at the two rectangles below. Which one has the greater perimeter?

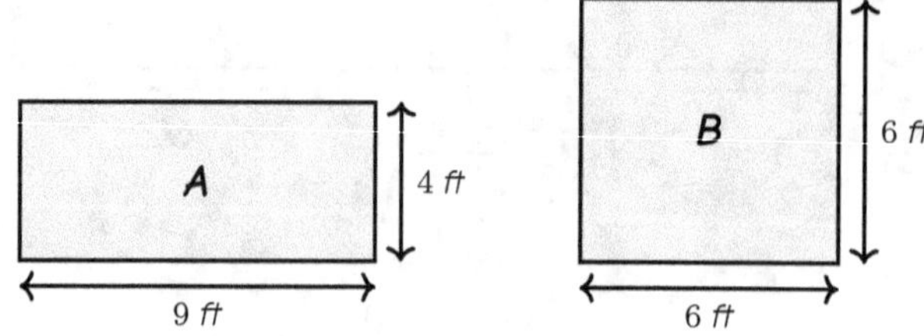

(A) Rectangle A (26 ft)

(B) Rectangle B (24 ft)

(C) They are equal

(D) Rectangle B (26 ft)

23. Lisa measures pencil lengths: 2 pencils at $\frac{5}{8}$ in, 3 pencils at $\frac{6}{8}$ in, 1 pencil at $\frac{7}{8}$ in. What is the longest pencil length measured?

(A) $\frac{5}{8}$ in

(B) $\frac{6}{8}$ in

(C) $\frac{7}{8}$ in

(D) $\frac{8}{8}$ in

24. A quarter turn (going one-fourth of the way around a circle) measures how many degrees?

(A) 45°

(B) 90°

(C) 180°

(D) 360°

25. Four angles meet at a point and together make a full 360° turn. Three of the angles measure 90°, 90°, and 80°. What is the fourth angle?

(A) 80°

(B) 90°

(C) 100°

(D) 110°

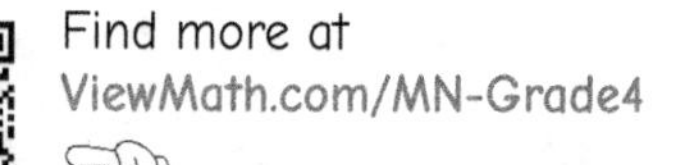

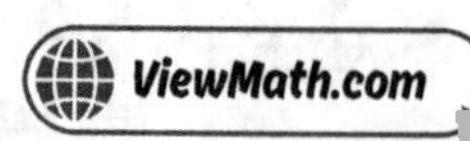

26. Maya draws a figure that has a beginning but no end. It goes on forever in one direction. What did she draw?

(A) A line

(B) A line segment

(C) A point

(D) A ray

27. Look at the angle below. Is it acute, right, or obtuse? Write your answer.

Your Answer:

28. Which of these real-world examples best shows perpendicular lines?

(A) Two lanes of a highway going straight ahead

(B) The top and side edges of a door

(C) Two train tracks running side by side

(D) The two sides of a ladder

29. "Every square is a rectangle, but not every rectangle is a square." This statement is:

(A) False — not every square is a rectangle

(B) True

(C) False — every rectangle is a square

(D) False — squares and rectangles are unrelated

30. How many lines of symmetry does an equilateral triangle have?

Your Answer:

 # End of Practice Test 2 

Great job finishing the test!

✅ My Score

I got _____________ out of 30 questions right.

*Check your answers in the **Answer Key** at the back of the book.*

💡 *Review any questions you missed. That's how we learn!*

📊 Check Your Score Online!

Visit **ViewMath Academy** to enter your answers and see which topics you need to review. You can also explore lessons, take quizzes, track your scores, and save your progress!

viewmath.com/score/4.1.MN.02

Or go to viewmath.com/score and enter code: 4.1.MN.02

3

Practice Test 3

 30 Questions

✏ Before You Start ✏

- ✔ **Read each question carefully** before choosing your answer.
- ✔ **Show your work** on scratch paper when you need to.
- ✔ **Skip hard questions** and come back to them later.
- ✔ **Check your answers** when you're done.
- ✔ **Take your time** — there's no rush!

 ⭐ You've Got This! ⭐

Do your best and show what you know!

1. The number bond below shows a multiplication comparison. What number belongs in the top circle?

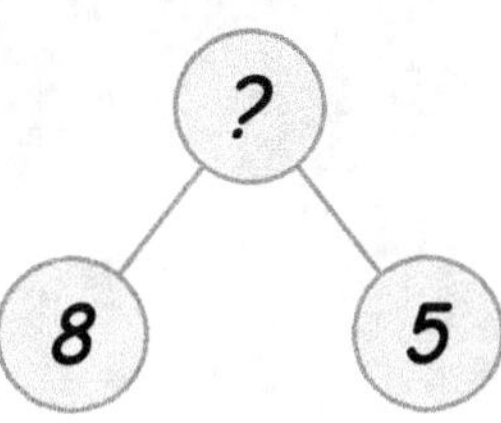

Hint: The top number is 8 times as many as 5.

Your Answer:

2. Sofia baked 48 cookies. She put them equally into 6 bags. She then bought 5 bags from a friend. How many bags does she have altogether?

(A) 11

(B) 8

(C) 13

(D) 43

3. Which statement is TRUE about all even numbers greater than 2?

(A) They are all prime.

(B) They are all composite.

(C) They are neither prime nor composite.

(D) Some are prime and some are composite.

4. Create your own number pattern using the rule "Add 8, starting at 4." Write the first five terms.

Your Answer:

5. Look at the base-ten blocks below. What number do they show?

(A) 374

(B) 347

(C) 743

(D) 437

6. A stadium holds 72,500 fans. A second stadium holds 72,050 fans. Which stadium holds fewer fans?

(A) First stadium

(B) Second stadium

(C) They hold equal fans

(D) Cannot determine

7. Look at the place value chart below. The digit 7 moves from one position to another.

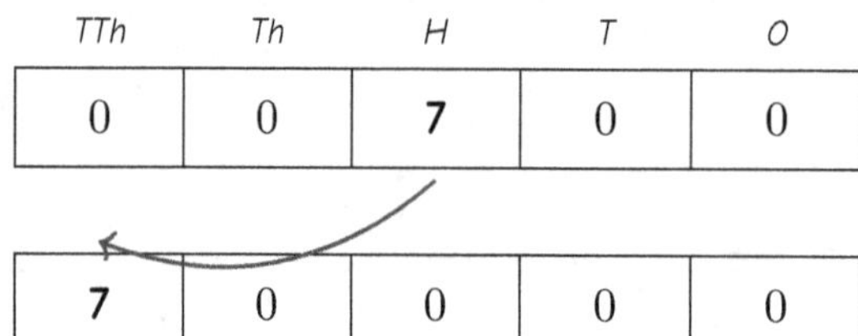

How did the value of the digit 7 change?

(A) It became 10 times bigger

(B) It became 100 times bigger

(C) It became 10 times smaller

(D) It stayed the same

8. Estimate $671 - 289$ by rounding each number to the nearest hundred.

Your Answer:

9. *What is* 7×905*?*

(A) 6,235

(B) 6,325

(C) 6,335

(D) 6,435

10. *63 books need to be packed into boxes that hold 8 books each. How many **full** boxes can be packed?*

(A) 6 boxes

(B) 7 boxes

(C) 8 boxes

(D) 9 boxes

11. $\frac{5}{6} = \frac{3}{6} + \frac{?}{6}$. *What is the missing numerator?*

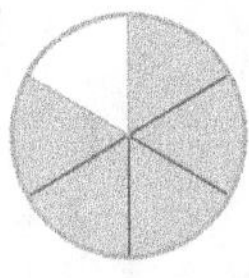

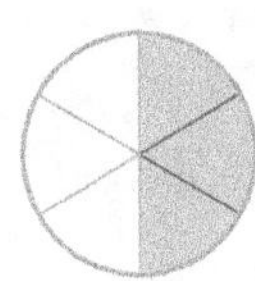

(A) 1

(B) 3

(C) 2

(D) 4

12. What is $\frac{3}{10} + \frac{3}{10}$?

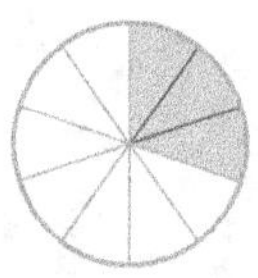

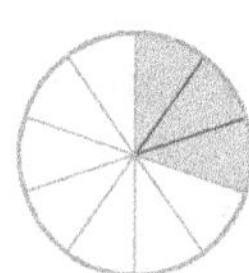

(A) $\frac{6}{20}$

(B) $\frac{9}{10}$

(C) $\frac{3}{10}$

(D) $\frac{6}{10}$

13. Convert $2\frac{3}{4}$ to an improper fraction.

(A) $\frac{5}{4}$

(B) $\frac{9}{4}$

(C) $\frac{11}{4}$

(D) $\frac{8}{4}$

14. What is $2 \times \frac{3}{4}$, written as a mixed number?

(A) $1\frac{2}{4}$

(B) $1\frac{1}{4}$

(C) $1\frac{3}{4}$

(D) $\frac{6}{4}$

15. The fraction bar below shows $\frac{3}{10}$. What is this fraction written with a denominator of 100?

(A) $\frac{3}{100}$

(B) $\frac{13}{100}$

(C) $\frac{30}{100}$

(D) $\frac{300}{100}$

16. Write 0.09 as a fraction.

Your Answer:

17. A store sells juice for $0.85 and water for $0.79. Which drink costs more? Write the comparison using >.

Your Answer:

18. How many centimeters are in 1 meter?

(A) 10

(B) 100

(C) 1,000

(D) 10,000

19. Convert 36 inches to feet.

(A) 2 ft

(B) 3 ft

(C) 4 ft

(D) 12 ft

20. A dog weighs 5 kg. A cat weighs 3,000 g. Which animal is heavier?

(A) The cat

(B) The dog

(C) They weigh the same

(D) Cannot be determined

Find more at
ViewMath.com/MN-Grade4

21. A room has the shape shown below with a rectangular piece cut out. Find the area of the shaded region.

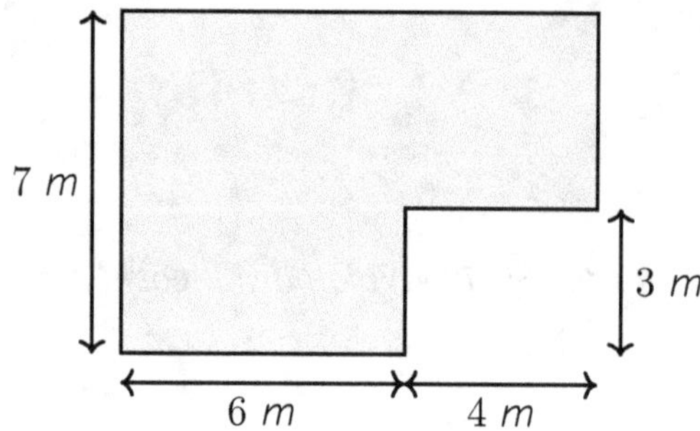

Your Answer:

22. A picture frame has outer dimensions 14 in by 11 in and inner dimensions 10 in by 7 in. What is the perimeter of the outer frame?

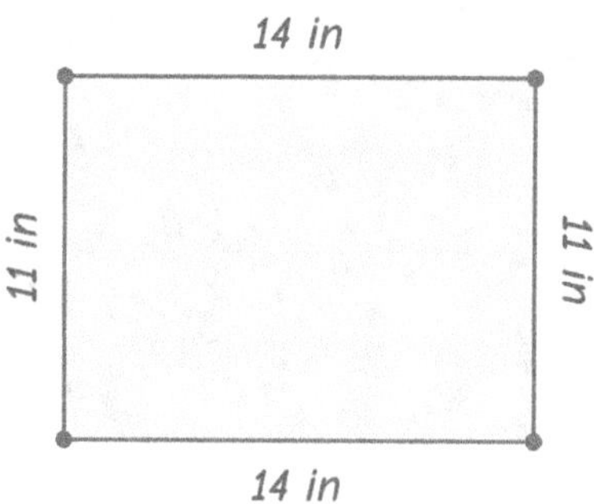

(A) 25 in

(B) 42 in

(C) 50 in

(D) 54 in

23. A line plot shows: 2 X's at $\frac{1}{4}$, 3 X's at $\frac{1}{2}$, and 1 X at $\frac{3}{4}$. How many total data points are there?

(A) 3

(B) 6

(C) 8

(D) 12

24. An angle that measures 55° is called a(n) ___________ angle.

Your Answer:

25. Two adjacent angles share a side: one is 65° and the other is 25°. What is the total angle?

Your Answer:

26. How many endpoints does a **line** have?

(A) 0

(B) 1

(C) 2

(D) 4

27. Lena says that an angle of 91° is a right angle because it is "close to 90°." Is she correct?

(A) Yes, because it is almost 90°

(B) No, it is an acute angle

(C) No, it is an obtuse angle

(D) No, it is a straight angle

28. Which pair of lines below is **perpendicular**?

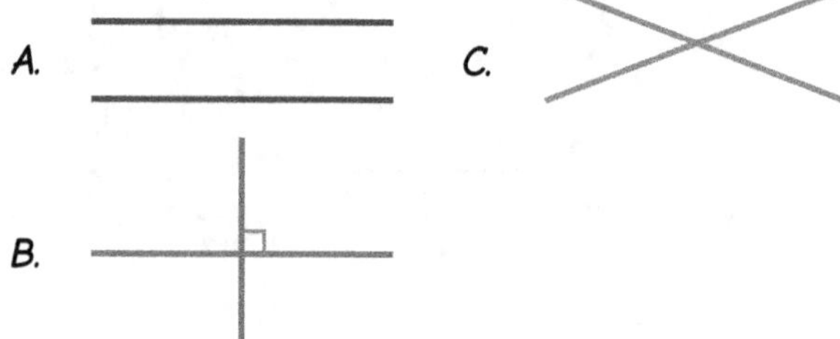

(A) Pair A

(B) Pair B

(C) Pair C

(D) Both Pair A and Pair C

29. A polygon with 5 sides is called a:

(A) Hexagon

(B) Octagon

(C) Pentagon

(D) Decagon

30. How many lines of symmetry does a square have?

Your Answer:

⭐ *End of Practice Test 3* ⭐

Great job finishing the test!

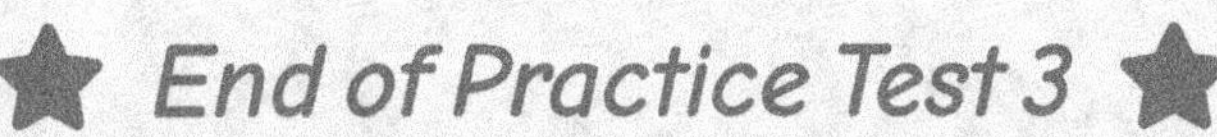

☑ My Score

I got _____________ out of 30 questions right.

*Check your answers in the **Answer Key** at the back of the book.*

💡 Review any questions you missed. That's how we learn!

📊 Check Your Score Online!

*Visit **ViewMath Academy** to enter your answers and see which topics you need to review. You can also explore lessons, take quizzes, track your scores, and save your progress!*

viewmath.com/score/4.1.MN.03

*Or go to **viewmath.com/score** and enter code: 4.1.MN.03*

Answer Key & Explanations

Check Your Answers!

First try each test on your own, then look here to check.

Read the explanations to learn from any mistakes ★

Practice Test 1 — Answer Key

1 35

2 15

3 Yes, 100 is a multiple of 4 ($4 \times 25 = 100$). No, 100 is not a multiple of 6 ($100 \div 6 = 16$ R4).

4 Rule: Multiply by 5; missing value = 45

5 C

6 C

7 B

8 B

9 2,996

10 B

11 C

12 C

13 C

14 C

15 A

16 B

17 B

18 A

19 B

20 1,200 m (or 1 km 200 m)

21 B

22 B

23 C

24 C

25 B

26 B

27 D

28 False

29 C

30 C

Time to Learn!

Go through the explanations below, **especially for the questions you missed**.

Understanding why each answer is correct makes you a stronger math thinker!

👍 **Tip:** Circle any questions you got wrong, then read their explanation carefully.

Practice Test 1 — Detailed Explanations

1. $7 \times 5 = 35$ pencils.

2. Step 1: $45 \div 5 = 9$ students per team (not needed). There are 5 teams each needing 3 jump ropes: $5 \times 3 = 15$.

3. $100 \div 4 = 25$ exactly; $100 \div 6$ leaves a remainder.

4. $3 \times 5 = 15$, $4 \times 5 = 20$, $6 \times 5 = 30$, so $9 \times 5 = 45$.

5. The hundred-thousands place is the leftmost digit of a six-digit number. In 307,416, the digit 3 is in the hundred-thousands place.

6. In 410,300 vs 410,030: compare hundreds: $3 > 0$. So $410,300 > 410,030$. The $<$ sign is wrong—this comparison is INCORRECT.

7. $3 \times 100 = 300$. Multiplying by 100 shifts all digits two places to the left (same as multiplying by 10 twice).

8. $387 \approx 400$ (tens digit $8 \geq 5$, round up) and $425 \approx 400$ (tens digit $2 < 5$, round down). $400 + 400 = 800$.

9. $7 \times 400 = 2{,}800$; $7 \times 20 = 140$; $7 \times 8 = 56$. Total: $2{,}800 + 140 + 56 = 2{,}996$.

10. $47 \div 5 = 9$ R2. You can buy 9 notebooks (costing \$45) and will have \$2 left over — not enough for another notebook.

11. 3 slices of size $\frac{1}{6}$ each: $\frac{1}{6} + \frac{1}{6} + \frac{1}{6} = \frac{3}{6}$.

12. Total gas used: $\frac{2}{10} + \frac{3}{10} = \frac{5}{10}$. Gas left: $\frac{9}{10} - \frac{5}{10} = \frac{4}{10}$.

13. Whole numbers: $3 + 2 = 5$. Fractions: $\frac{4}{5} + \frac{3}{5} = \frac{7}{5} = 1\frac{2}{5}$ (regroup). Add: $5 + 1\frac{2}{5} = 6\frac{2}{5}$.

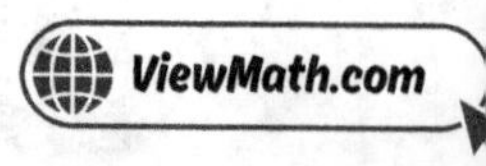

14 $5 \times 2 = 10$; $\frac{10}{8} = 1\frac{2}{8}$ because $10 \div 8 = 1$ R 2.

15 You must convert to the same denominator first: $\frac{1}{10} = \frac{10}{100}$. Then $\frac{10}{100} + \frac{3}{100} = \frac{13}{100}$.

16 $0.6 = \frac{6}{10} = \frac{60}{100}$ (multiply top and bottom by 10).

17 Both have 0 ones. Compare tenths: $3 < 4$, so $0.34 < 0.43$.

18 $1\ cm = 10\ mm$. The other conversions are wrong: $1\ m = 100\ cm$, $1\ km = 1,000\ m$, $1\ kg = 1,000\ g$.

19 There are 12 inches in 1 foot.

20 3 laps $\times 400\ m = 1,200\ m$. Since $1,000\ m = 1\ km$, the runner has covered 1 km and 200 m.

21 Large rectangle: $8 \times 6 = 48$ sq m. Cut out: $2 \times 2 = 4$ sq m. Remaining: $48 - 4 = 44$ sq m.

22 $36 = 2(10) + 2w \Rightarrow 16 = 2w \Rightarrow w = 8\ cm.$

23 When measuring in halves, the only value between 0 and 1 is $\frac{1}{2}$.

24 An obtuse angle is between $90°$ and $180°$. Only $110°$ fits that range.

25 $45 + 45 = 90°$, which is a **right** angle.

26 All three points are on the same line, so $\overline{CD}$ and $\overline{DE}$ are both parts (segments) of that one line.

27 In order from smallest to largest: acute $(< 90°)$, right $(= 90°)$, obtuse $(< 180°)$, straight $(= 180°)$. Straight is the largest.

28 Parallel lines never intersect, no matter how far they are extended.

29 A rectangle always has four right angles. Its opposite sides are equal and parallel (making it a parallelogram).

30 An equilateral triangle has 3 lines of symmetry — one from each vertex to the midpoint of the opposite side.

✅ Practice Test 2 — Answer Key

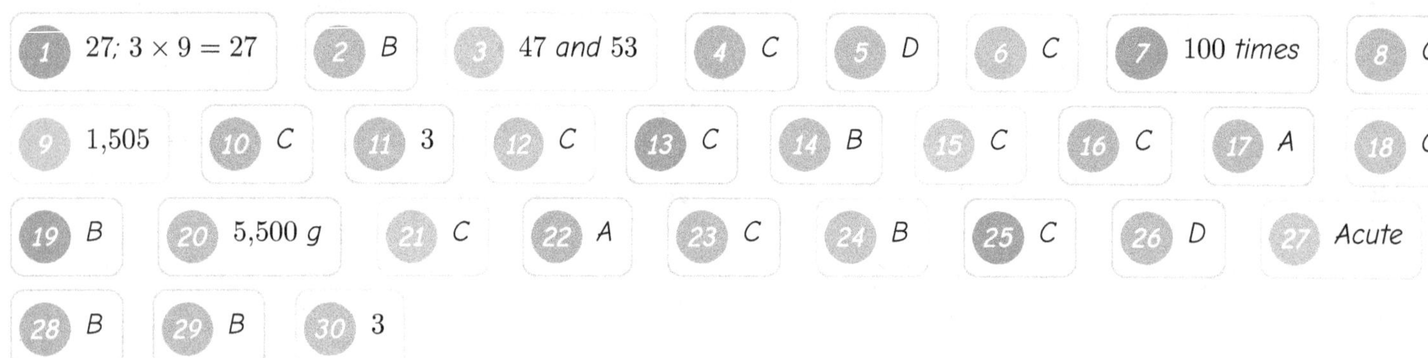

1	$27; 3 \times 9 = 27$	2	B	3	47 and 53	4	C	5	D	6	C	7	100 times	8	C
9	1,505	10	C	11	3	12	C	13	C	14	B	15	C	16	C	17	A	18	C
19	B	20	5,500 g	21	C	22	A	23	C	24	B	25	C	26	D	27	Acute
28	B	29	B	30	3

💡 Time to Learn! 💡

Go through the explanations below, **especially for the questions you missed**.

Understanding why each answer is correct makes you a stronger math thinker!

👍 **Tip:** Circle any questions you got wrong, then read their explanation carefully.

📖 Practice Test 2 — Detailed Explanations

1 The array has 3 rows and 9 columns. $3 \times 9 = 27$ dots total.

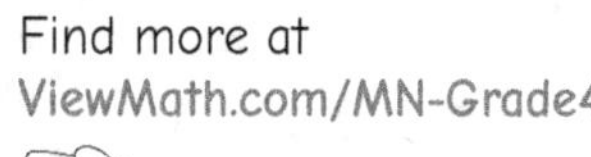
Find more at
ViewMath.com/MN-Grade4

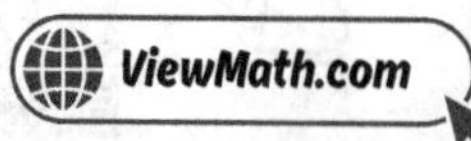

2   Step 1: $7 \times 24 = 168$ crayons. Step 2: $168 - 15 = 153$ crayons.

3   47 has only factors 1 and 47. 53 has only factors 1 and 53. All other numbers between 45 and 55 are composite.

4 $3 \times 3 = 9$ (2nd), $9 \times 3 = 27$ (3rd), $27 \times 3 = 81$ (4th).

5   In 850,293: 8 = hundred-thousands, 5 = ten-thousands. The value is $5 \times 10,000 = 50,000$.

6   Both towns have 183 in the thousands period. Compare hundreds: $4 > 0$. So $183,400 > 183,040$. Town A has more people.

7   Position A is the ten-thousands place ($9 \times 10,000 = 90,000$). Position B is the hundreds place ($9 \times 100 = 900$). $90,000 \div 900 = 100$. Position A is 100 times bigger.

8   The tens digit is 4. Look right at the ones digit: $7 \geq 5$, so round up. 4 becomes 5, giving 350.

9 7×215: ones $7 \times 5 = 35$, write 5, carry 3; tens $7 \times 1 + 3 = 10$, write 0, carry 1; hundreds $7 \times 2 + 1 = 15$. Product is 1,505.

10 $7 \times 72 = 504$. Check: $7 \times 70 = 490$ and $7 \times 2 = 14$; $490 + 14 = 504$.

11 $? + 2 + 1 = 6$, so $? = 3$.

12 $6 - 2 = 4$; denominator stays 8: $\frac{4}{8}$.

13 $(3 \times 5) + 2 = 17$, so $3\frac{2}{5} = \frac{17}{5}$.

14 When multiplying a fraction by a whole number, multiply ONLY the numerator. The denominator stays the same. The correct answer is $\frac{6}{5}$.

15 Convert $\frac{3}{10} = \frac{30}{100}$. Then $\frac{30}{100} + \frac{5}{100} = \frac{35}{100}$.

16 $0.30 = \frac{30}{100} = \frac{3}{10} = 0.3$. Adding a zero at the END of a decimal does not change its value.

17 Grid A has 42 squares shaded and Grid B has 38 squares shaded. Since $42 > 38$, $0.42 > 0.38$.

18 Step 1: 2 km $= 2 \times 1,000 = 2,000$ m. Step 2: 2,000 m $= 2,000 \times 100 = 200,000$ cm.

19 1 lb $= 16$ oz. So 5 lb $= 5 \times 16 = 80$ oz.

20 Convert: 4 kg $= 4,000$ g. Then $4,000 + 1,500 = 5,500$ g.

21 $l = 108 \div 9 = 12$ ft.

22 A: $P = 2 \times (9 + 4) = 26$ ft. B: $P = 2 \times (6 + 6) = 24$ ft. Rectangle A has the greater perimeter.

23 $\frac{7}{8}$ inch is the largest value on the line plot.

24 One-quarter of $360°$ is $360 \div 4 = 90°$. A quarter turn makes a right angle.

25 $90 + 90 + 80 = 260$. Fourth angle $= 360 - 260 = 100°$.

26 A ray has one endpoint (a beginning) and goes on forever in one direction.

27 The angle opens less than $90°$ (less than a right angle), so it is acute.

Find more at
ViewMath.com/MN-Grade4

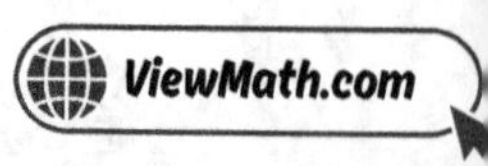

28 · The top edge and side edge of a door meet at a 90° angle, making them perpendicular.

29 · A square has all properties of a rectangle (4 right angles, opposite equal sides), so every square IS a rectangle. But rectangles don't need equal sides, so not every rectangle is a square.

30 · An equilateral triangle has 3 lines of symmetry, one from each vertex to the midpoint of the opposite side.

✅ Practice Test 3 — Answer Key

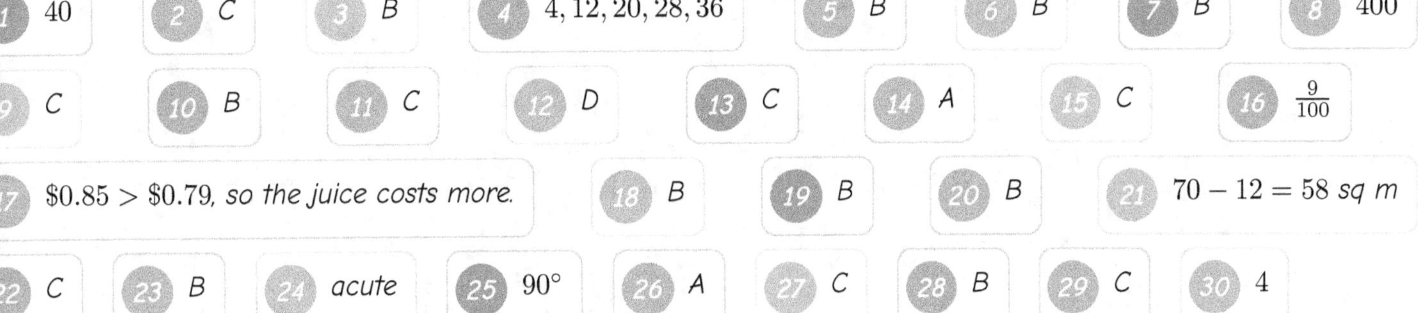

1 · 40	2 · C	3 · B	4 · 4, 12, 20, 28, 36	5 · B	6 · B	7 · B	8 · 400
9 · C	10 · B	11 · C	12 · D	13 · C	14 · A	15 · C	16 · $\frac{9}{100}$
17 · $0.85 > $0.79, so the juice costs more.	18 · B	19 · B	20 · B	21 · $70 - 12 = 58$ sq m
22 · C	23 · B	24 · acute	25 · 90°	26 · A	27 · C	28 · B	29 · C	30 · 4

💡 Time to Learn! 💡

Go through the explanations below, **especially for the questions you missed.**

Understanding why each answer is correct makes you a stronger math thinker!

👍 **Tip:** Circle any questions you got wrong, then read their explanation carefully.

📖 Practice Test 3 — Detailed Explanations

1 · 8 times as many as 5 is $8 \times 5 = 40$.

Find more at
ViewMath.com/MN-Grade4

2 Step 1: $48 \div 6 = 8$ bags baked. Step 2: $8 + 5 = 13$ bags total.

3 Every even number greater than 2 has 2 as a factor (in addition to 1 and itself), giving more than 2 factors, so all are composite.

4 $4 + 8 = 12$, $12 + 8 = 20$, $20 + 8 = 28$, $28 + 8 = 36$.

5 There are 3 hundreds (300), 4 tens (40), and 7 ones (7). That makes $300 + 40 + 7 = 347$.

6 Both start with 72. Compare hundreds: $5 > 0$. So $72{,}500 > 72{,}050$. The second stadium holds fewer fans.

7 The 7 moved from the hundreds place (700) to the ten-thousands place (70,000). That's two places to the left: $700 \times 10 \times 10 = 70{,}000$, which is 100 times bigger.

8 $671 \approx 700$ (tens digit $7 \geq 5$, round up) and $289 \approx 300$ (tens digit $8 \geq 5$, round up). $700 - 300 = 400$.

9 Ones: $7 \times 5 = 35$, write 5, carry 3. Tens: $7 \times 0 + 3 = 3$. Hundreds: $7 \times 9 = 63$. Product is $6{,}335$.

10 $63 \div 8 = 7$ R7. There are 7 full boxes (holding 56 books), with 7 books left over. Since those 7 remaining books do not fill another complete box of 8, you can only fill 7 complete boxes.

11 $3 + ? = 5$, so $? = 2$. The missing numerator is 2.

12 $3 + 3 = 6$; keep denominator 10: $\frac{6}{10}$.

13 Multiply the whole number by the denominator and add the numerator. $(2 \times 4) + 3 = 11$, so $\frac{11}{4}$.

14 $2 \times 3 = 6$; $\frac{6}{4} = 1\frac{2}{4}$ because $6 \div 4 = 1$ R 2.

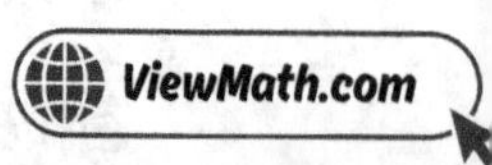

15 $\frac{3}{10} = \frac{3\times10}{10\times10} = \frac{30}{100}$. *Multiply top and bottom by 10.*

16 *0.09 has a 9 in the hundredths place, so it equals* $\frac{9}{100}$.

17 *Compare tenths digits: 0.85 has 8 in the tenths place; 0.79 has 7 in the tenths place. Since $8 > 7$, $0.85 > 0.79$. Juice costs more.*

18 *There are 100 centimeters in 1 meter.*

19 $36 \div 12 = 3$ *ft.*

20 *Convert:* $5\ kg = 5{,}000\ g$. *Since* $5{,}000 > 3{,}000$, *the dog is heavier.*

21 *Full rectangle:* $10 \times 7 = 70$ *sq m. Cut-out:* $4 \times 3 = 12$ *sq m. Shaded area* $= 70 - 12 = 58$ *sq m.*

22 *Outer perimeter:* $P = 2(14 + 11) = 2 \times 25 = 50$ *in.*

23 $2 + 3 + 1 = 6$ *total data points.*

24 $55°$ *is less than* $90°$, *so it is an acute angle.*

25 $65 + 25 = 90°$.

26 *A line goes on forever in both directions and has no endpoints at all.*

27 *A right angle is EXACTLY* $90°$. *Since* $91°$ *is slightly more than* $90°$, *it is obtuse.*

28 *Perpendicular lines meet at a $90°$ angle (shown by the small square). Only Pair B has a right angle mark.*

29 A pentagon has 5 sides (penta = five).

30 A square has 4 lines of symmetry: one vertical, one horizontal, and two diagonal.

Great job checking your work!

Keep practicing and you'll be a math star!

www.ingramcontent.com/pod-product-compliance
Lightning Source LLC
Chambersburg PA
CBHW060207120726
48004CB00007B/1724